Geothermal Heat pumps What You Need to Know Before You Buy One

Jerry Scherer and

Miss Jacquelyn Scherer

DEDICATION

This book is dedicated to all who are striving for a greener earth.

Acknowledgments

My hat is off to all of my DIY geothermal heat pump customers who have been able to successfully install their geothermal heat pump systems. The testing information that they gathered for me over the past 10 years has proven that they did a better job installing their geothermal heat pump system then most geothermal heat pump contractors do.

TABLE OF CONTENTS

Page 8   Preface

Page 11   **Chapter One**

Geothermal heat pump tax credits.

Page 13   **Chapter Two**

About geothermal heat pumps.

Page 14   **Chapter Three**

The three loops in a geothermal heat pump system.

Page 16   **Chapter Four**

The air has problems, The earth has solutions.

Page 18   **Chapter Five**

How do heat pumps move heat from cooler areas into warmer areas?

Page 20   **Chapter Six**

Any geothermal heat pump is better than an air source heat pump.

Page 24   **Chapter Seven**

What do geothermal heat pumps have to do with renewable energy?

Page 26   **Chapter Eight**

Will a geothermal heat pump keep me comfortable? YOU BET IT WILL!!!

Page 27   **Chapter Nine**

Geothermal heat pumps: Keeping maintenance simple.

Page 28   **Chapter Ten**

Which efficiency ratings should I compare? SEER, EER, COP?

Page 29   **Chapter Eleven**

Efficiency: Geothermal heat pumps compared to electric resistant heaters

Page 30   **Chapter Twelve**

Does running a ceiling fan lower my cooling bills?

Page 32   **Chapter Thirteen**

Can ceiling fans lower my heating bills?

Page 34   **Chapter Fourteen**

Properly sizing your geothermal heat pump will make a big difference.

Page 36   **Chapter Fifteen**

Rules of thumb for sizing are out!!!

Page 38   **Chapter Sixteen**

Closed earth loops.

Page 40   **Chapter Seventeen**

Horizontal earth loops.

Page 48   **Chapter Eighteen**

Slinky loops.

Page 49   **Chapter Nineteen**

Naeem's slinky earth loop installation.

Page 72   **Chapter Twenty**

Pond loops.

Page 73   **Chapter Twenty-one**

Vertical bore loops.

Page 80   **Chapter Twenty-two**

Using well water for an open loop.

Page 84   **Chapter Twenty-three**

Geothermal heat pump earth loop antifreeze.

Page 89   **Chapter Twenty-four**

What is insulation R-value?

Page 90   **Chapter Twenty-five**

Best insulation R-value per dollar:

Stop the largest losses/leaks first.

Page 94   **Chapter Twenty-six**

What is a desuperheater?

Page 96   **Chapter Twenty-seven**

Single stage, Two stage, and

Variable speed geothermal heat pumps.

Page 99

Chapter Twenty-eight

Water-to-Water VS. Water to Air

Geothermal heat pumps.

Page 102

Chapter Twenty-nine

Geothermal heat pumps:

Packaged units VS. Split systems.

Page 104

Chapter Thirty

ClimateMaster's

ClimaDry geothermal heat pump

dehumidification system.

Page 106

Chapter Thirty-one

Geothermal heat pump thermostat:

Don't set it back.

Page 109

Chapter Thirty-two

Duel Fuel geothermal heat pump systems.

Page 111   **Chapter Thirty-three**

About heat distribution.

Page 112   **Chapter Thirty-four**

Sizing and designing your forced air ducting system (duct work).

Page 117   **Chapter Thirty-five**

Sizing and designing your

Geothermal heat pump.

Page 119   **Chapter Thirty-six**

Designing your closed earth loop.

Page 123   **Chapter Thirty-seven**

What we need from you so we can size and design your horizontal earth loop.

Page 125   **Chapter Thirty-eight**

How to test your soil.

Page 129   **About the Author**

Preface

The purpose of this geothermal heat pump guide is to present you with the information that you will need so you will be able to get a successfully designed and installed geothermal heat pump heating and cooling system for your home. We will give you the information needed so you will be able to either have a geothermal heat pump contractor install your system or, if you are a DIYer, you will know the correct steps to take so you can install it yourself.

The reason we decided to write this geothermal heat pump guide is because of the need for the general public to understand the steps involved with geothermal heat pump installations.

We started selling geothermal heat pumps nationwide 12 years ago. Our prospective customers were people who had already gotten one or more estimates. After they picked themselves up off of the floor because of the sticker shock, they started looking for information on the web about geothermal heat pumps and they found us. As we talked to our prospective customers we realized that there was a tremendous problem in the geothermal heat pump industry. As we asked these people questions about their estimates they told us a frightening tale that was almost inconceivable to us.

We learned that the geothermal heat pump industry is most contractors don't know what they are doing. We were amazed because the industry requires their contractors to be certified, and they must take classes to get the certification.

We also learned the far majority of geothermal heat pump contractors are installing over-sized geothermal heat pumps, and under-sized earth loops, and the combination of the two will cause your geothermal heat pump to operate inefficiently. This kind of operation will save you little to no money on your heating and cooling bills. It will also cause the geothermal heat pump to short cycle, and this short cycling will reduce its life. The amount the geothermal heat pump has been over-sized and the amount the earth loop has been under-sized will determine how inefficient it will be and how much the geothermal heat pump's life has been shortened.

It is because of this that we felt people should be made aware of the proper step by step procedure so the homeowner can be aware of an unqualified geothermal heat pump contractor, and thereby save themselves a lot of money, and unimaginable trouble.

Nothing compares to the money that you can save, the comfort you can have, and the trouble-free operation of a correctly designed & installed geothermal heat pump system.

BUT REMEMBER THIS!!!!!!!!!!!!!!

Nothing compares to the trouble you can have with an incorrectly designed and installed geothermal heat pump system.

GEOTHERMAL HEAT PUMP TAX CREDIT

Federal tax credits are available for residential geothermal heat pump system installations! Under the Residential Renewable Energy Tax Credit, 30% of the cost of residential geothermal heat pump installations may be claimed, with no upper limit. A Tax Credit is $1.00 credit for $1.00 owed (whereas a deduction just removes a percentage of the tax owed). The tax credit can be carried over from year to year if it is not all used within the same year it is credited.

In order to qualify for the tax credit, the geothermal heat pump must meet or exceed the Energy Star Program requirements (for example EER and COP) that were effective when the geothermal heat pump was installed. You can claim everything that it cost you to upgrade your heating and cooling system to the geothermal heat pump system - the cost of the geothermal heat pump, the earth loop piping, all of the parts and materials for the system, all of the digging costs (if you pay to have it done), and any labor you have to pay to have it installed.

Starting December 1, 2009, the geothermal heat pump does not have to have water heating (hot water generator (HWG), desuperheater) in order to qualify for the tax credit.

Please see the following links to the Department of Energy and the Energy Star websites for more information:

[Energy Star Tax Credits](#)

[Department of Energy Tax Breaks](#)

The federal tax credit expires December 31, 2016, so start your geothermal heat pump installation today!

ABOUT GEOTHERMAL HEAT PUMPS

A geothermal heat pump is a heating and cooling machine. Geothermal heat pumps heat and cool your home by moving heat between the earth and your home. A geothermal heat pump does the same thing as an air-source heat pump, except it uses the earth instead of the air for heat transfer("geo" means earth, and "thermal" means heat). The heat that it transfers into the home in the winter is heat that the earth has absorbed from the summer sunshine. Since you already own this heat, it's free for the taking you only pay to move it inside the house. A geothermal heat pump is the only way to gather up and use this free source of heat.

Geothermal heat pumps can be the most energy efficient (and environmentally clean) way to heat and cool your home, but they must be matched closely to your home, and the earth it is coupled to, to achieve this. Because every homesite is unique, there are no rules of thumb for the design of a geothermal heat pump system.

THE 3 LOOPS IN A GEOTHERMAL HEAT PUMP SYSTEM

To move heat between the rooms of your home and the earth, a geothermal heat pump system uses 3 types of loops, and 2 heat exchangers.

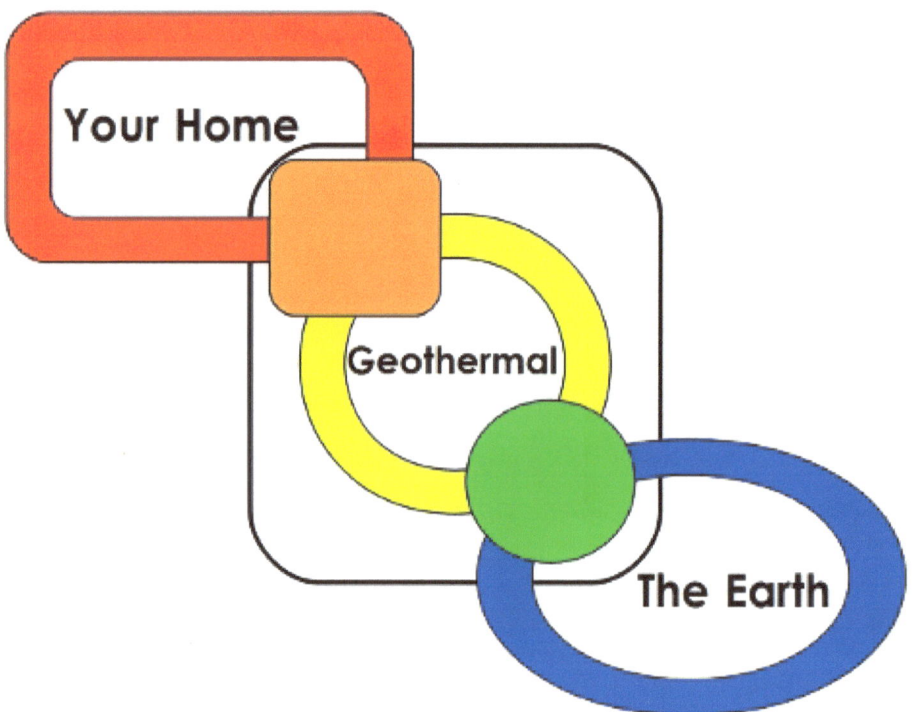

1. The earth loop moves heat between the earth, and the geothermal heat pump's water-to-refrigerant/Source-side coil.

2. Inside the geothermal heat pump, the refrigerant loop moves heat between the two heat exchanger coils: the water-to-refrigerant/Source-side coil, and the air-to-refrigerant coil or the water-to-refrigerant/Load-side coil.

3. The air ducting loop, or the hydronic piping loop, moves heat between the air-to-refrigerant coil, or the water-to-refrigerant/Load-side coil, and the rooms of your home.

THE AIR HAS PROBLEMS, THE EARTH HAS SOLUTIONS

Using the earth, instead of the air, eliminates the problems that happen to air-source heat pumps when the temperature goes below freezing. Unfortunately, below about 35° F, the air is changed in several ways that make very little heat available for air-source heat pumps to use. The combination of frost build-up, and much drier air (holding much less heat), prevents an air-source heat pump from being able to transfer enough heat to keep your home warm. And there are a lot of places that go below 35° F in the winter. Must we abandon heat pumps then?

No, there is another avenue for success! The ground 4 to 6 feet below us doesn't get that cold. It stays around 40 to 50° F all year.* Since heat pumps work extremely well at 35 to 60° F, installing the heat exchanger in the earth where the temperature maintains 40 degrees, even in the middle of winter, means geothermal (ground-source) heat pumps completely avoid the problems of air-source heat pumps, and allow us to take advantage of all the free heat in the earth!

*Equally important is that ice that coats an earth-surrounded heat exchanger doesn't slow the movement of heat into it, while ice that coats an air-surrounded heat exchanger severely limits the amount of heat that can move into it. This is because gases (the air) have much lower conduction coefficients than solids do.

HOW DO HEAT PUMPS MOVE HEAT FROM COOLER AREAS INTO WARMER AREAS?

At first the thing that comes into our mind is, "The earth is cool - there's no heat in it; it's not warm enough to heat my home!" Although the earth is a lower temperature than we'd like our homes to be, it still has heat in it. Heat normally moves from warmer areas into colder ones, but using the same principle that makes air conditioners and refrigerators work, we can transfer the heat from the cooler earth into your warmer home in the winter.

An air conditioner turns hot air into cold air by removing the heat and pushing it outside. In the summer, the inside air temperature of your home may be 75° F, and the outside air temperature of your home may be 90° F. With an air conditioner, we take heat from the lower inside temperature of your home, and concentrate/condense it into higher temperature heat - higher than the outside air temperature. That makes the heat easy to move into the outside air. An air conditioner/air-source heat pump is thus able to move heat from a lower temperature into a higher one.

Since a heat pump can move heat from any cooler area into a warmer one, we can use it in the winter also: we just turn it around/reverse it, and take heat out of the cooler outside air or earth, and move it into the warmer house.

ANY GEOTHERMAL HEAT PUMP IS BETTER THAN AN AIR-SOURCE HEAT PUMP

Why is a geothermal heat pump better than an air-source heat pump?

Water holds a huge amount of heat, compared to most substances. You can take 62.42 BTUs out of 1 cubic foot of water (at 40° F), and its temperature will drop just 1 degree. In contrast, air that is dry, or almost dry (which it will be at 40° F), holds very little heat. You can only take 0.019 BTUs out of 1 cubic foot of air before its temperature drops 1 degree. So, comparing equal volumes, water holds 3274.78 times as much heat as dry air does!

What does this mean for heat pumps? An air-source heat pump must move about 3000 times the volume of air through its heat exchanger, than a geothermal heat pump does water, to get the same heat. Even though air is less viscous than water, and easier to move, the air-source heat pump still ends up working harder: to move that much air, an air-source heat pump must run a fan consuming 600 watts of power per hour. A geothermal heat pump's water circulating pump only consumes 90 watts per hour!

Air has another problem for air-source heat pumps: the drier it is, the less heat it can hold, and colder air is drier than warmer air. So, as the temperature drops linearly, the amount of heat in the air drops exponentially! At about 34° F, almost all of the water vapor left in the air will frost out of it, which drastically lowers the heat available to the heat pump. The only way to compensate is to speed up the heat pump's fan, and move even larger quantities of air but this really ratchets up the power consumption. The combination of transferring less heat, and using more energy to do it, is crippling to the air-source heat pump's efficiency.

But water vapor isn't through causing trouble yet. When all that moisture starts frosting out of the air, it freezes right onto the heat exchanger coil of the air-source heat pump. Eventually the ice gets so thick that the fan can't pull any air through the heat exchanger coil, and the heat pump stops working. To solve this problem, the air-source heat pump has to defrost the outdoor coil several times per hour - and it takes quite a lot of power to do it. A geothermal heat pump, on the other hand, never requires defrosting (antifreeze keeps the loop fluid flowing).

Finally, the most important reason a geothermal heat pump is better than an air-source heat pump, is that the earth has more stable temperatures than the air. This gives a geothermal heat pump a much higher entering fluid temperature in the heating season, and a much lower entering fluid temperature in the cooling season, than an air-source heat pump, making the geothermal heat pump tremendously more efficient, with a much longer lifetime.

HEATING COMPARISON:
JANUARY IN GRAND RAPIDS, MINNESOTA @ 0° F

A 3-ton Amana ASZ air-source heat pump will have 14,700 BTUHs of heating capacity (1.23 tons), and a COP of 2.1.

A 3-ton ClimateMaster TTV geothermal heat pump (entering fluid temperature of 32° F) will have 29,000 BTUHs of heating capacity (2.42 tons), and a COP of 4.0.

The geothermal heat pump gives you twice as many BTUHs, at half the cost per BTUH!

COOLING COMPARISON:
AUGUST IN HOUSTON, TEXAS @ 105° F

The 3-ton Amana ASZ air-source heat pump will have 33,500 BTUHs of cooling capacity (2.79 tons), and an EER of 11.84 (it only gets two-thirds of that EER and capacity if it's covered with grass clippings and dirt, which obstruct the airflow).

The 3-ton ClimateMaster TTV geothermal heat pump (entering fluid temperature of 77° F) will have 38,200 BTUHs of cooling capacity (3.18 tons), and an EER of 18.2 (mowing the lawn won't hurt a geothermal heat pump's efficiency or capacity).

In all cases, a geothermal heat pump beats the pants off an air-source heat pump in fact, no matter what air-source heat pump you compare it to, ANY geothermal heat pump will perform better!

WHAT DO GEOTHERMAL HEAT PUMPS HAVE TO DO WITH RENEWABLE ENERGY?

The renewable energy source is **heat** - heat that is stored in the earth! It is amazing that we can take heat from 40 to 55 degree F dirt and heat our houses toasty warm. Geothermal heat pumps move heat out of the ground and into the house in the winter, and back into the ground from inside our houses in the summer. Instead of burning something to make new heat for our homes, we take heat that's already there and reuse it. It's renewable energy!

The earth's surface is kept warm by the sun's energy being absorbed into it. This heat can be removed from the earth and used to heat your home. Although there are many skeptics, the proof is plentiful. With a little help from a geothermal heat pump, we can move the heat from the earth into your home. A geothermal heat pump gathers up the earth's heat and squeezes it into a smaller 'ball' until it is hot enough to heat your home. It is so good at compressing the heat and moving it into your home, it can move 4 times as much heat as the energy it takes to move it.

And since the earth is cooler than your home in the summer it is a breeze to cool it with the earth's cool temperature. It is done by reversing the process: gathering up heat from the building and putting it back into the earth.

WILL A GEOTHERMAL HEAT PUMP KEEP ME COMFORTABLE? YOU BET IT WILL!

What! I'm just warming my butt....really.

Fossil fuel systems cause people to experience cold air drafts (as the warmer air rises). Geothermal heat pump systems run for longer periods of time causing the air to be moved more frequently and consistently. This increased air flow mixes all the room air resulting in a more comfortable and consistent temperature. Geothermal heat pump customers who have systems installed correctly always comment on being more comfortable.

When a geothermal heat pump system is planned, the duct work registers are set up so the air will wash the exterior walls with the heat and not allow the cold air to get to the floor. This continuous washing of the walls with warm air, makes the room temperature consistent. With a geothermal heat pump system, you do not experience cold drafts near the floor and your feet!

GEOTHERMAL HEAT PUMPS:

KEEPING MAINTENANCE SIMPLE

These type of systems are easy to take care of when the equipment is sized and installed correctly. Typically, changing the air filter will be the main effort needed to maintain your residential geothermal heat pump system, but as with any heating& cooling system a yearly checkup is a smart way to be sure everything is operating up to par. An inefficient system can quickly waste hundreds of dollars a year so the cost of a checkup could prove to be well worth it.

Like other equipment involving an air filter, a dirty filter will cause poor performance and lower efficiency. If a filter gets so dirty that airflow is severely restricted, the geothermal heat pump will protect itself by shutting down to prevent too high or low operating pressures. Most filters don't get to that point, but inefficiency isn't a desired outcome either, so remember to change your filters!

WHICH EFFICIENCY RATINGS SHOULD I COMPARE? SEER, EER, COP?

When air source heat pumps and air conditioners are rated they are rated with a SEER number. SEER is Seasonally adjusted Energy Efficiency Ratio. This represents the average efficiency of the unit over one 'season'. But, who will actually get this SEER from the unit? People in Miami, Florida or people in International Falls, Minnesota? The problem with the SEER is that it is not the true number for most homes. SEER ratings are ONLY good for comparing equipment that use SEER ratings, and it doesn't have much value otherwise.

The efficiency of a geothermal heat pump (EER: energy efficiency ratio and COP: coefficient of performance) is rated at specific entering water and air temperatures to the unit. So, if you are to compare apples to apples you will need to compare the EER and the COP of each geothermal heat pump calculated at the same rating conditions. However, geothermal heat pumps are so much more efficient than any other type of heating or cooling equipment, that any geothermal heat pump (properly sized and properly installed) will save you a tremendous amount of money.

EFFICIENCY: GEOTHERMAL HEAT PUMPS COMPARED TO ELECTRIC RESISTANT HEATERS

If we burn 1000 watts of electricity in any electric resistant heater ("space heaters", electric convection heater, or electric fan heater), we will get 3,410 BTU's of heat for the electricity we've paid for. We get all the heat out of the electricity we've paid for, so our efficiency is considered to be 1, or 100%.

Now, if we burn 1000 watts of electricity with a geothermal heat pump, we get about 4 times the heat, or 13,640 BTU's. We burn (and pay for) one unit of electricity and transfer 3 units (free) of heat. We get 4 units of heat for the price of 1!! This means our efficiency rating (COP: coefficient of performance) is 4, or 400 percent.

Geothermal Heat Pump efficiency is much greater than that of electric resistance heat!

DOES RUNNING A CEILING FAN LOWER MY COOLING BILLS?

One day several summers ago, we were called out to a family's home to take a look at why their air conditioning was not cooling their home, and what was causing their energy use to be much higher than normal. Upon entering their home we noticed that they had a ceiling fan in every room and they were all running at the same time. Next, their air conditioning system was checked and it was verified that the operation of it was correct. We asked if their A/C system always worked poorly, when the temperature got to the upper nineties, and they said "no" it had just been recently. We asked them if they always had ceiling fans in every room, and they said "no" (only one in the kitchen). They said they had read that ceiling fans would save them money on their cooling bills, so they had them installed.

Almost everyone has heard that ceiling fans will save money on cooling bills. When the air around us is moving, we usually feel cooler. This is because moving air evaporates our sweat faster than still air. Evaporating sweat cools us down faster than anything else, even slightly cooler air temperatures.

However, running a ceiling fan actually causes an air conditioned room to heat back up faster. The surfaces of the ceiling, exterior walls, and windows are warmer than the rest of the room. When you turn on a ceiling fan, it circulates the air in the room. This mixes the air in the room faster, which makes the air cooler next to the walls and ceiling, which in turn makes heat move faster through the walls into the room from outside. Ceiling fans only cool people down, not rooms, so if the fan is on and no one is in the room you are wasting electricity.

If your A/C is always on, then using ceiling fans will save you money if you use them only when you are in the room, and set your house thermostat to a slightly higher temperature. If you forget and leave them on when you leave the room, you won't save money. Running ceiling fans will cost you more money because your A/C unit will have to do more cooling, and the electricity burned by the fan will be completely wasted.

The reason that this family's energy use was so much higher, was that they had fans installed in every room of their home, and they had every one of them on continuously. Heat from outside was transferred into their home much faster, and it was so much faster that the A/C system couldn't even keep up. There was nothing at all the matter with the air conditioning system - the ceiling fans caused the higher energy use.

CAN CEILING FANS LOWER MY HEATING BILLS?

Ceiling fans can often be used to save money during the heating season. Most older homes have heating systems that are two to four times larger than they need to be. They also have blowers that operate at lower speeds. Being over sized will cause the system to blow hotter air than it should, and run very short cycles. When the system blows hotter air, the hotter air tends to stay up at the ceiling. The lower blower speeds also contribute to this. Also, when furnace cycles are shorter, there are longer periods without any air circulation. If the air is not moving in a room and it is cold outside, the cold air that is at the wall, and windows will slide down off of them onto the floor. This will make the temperature on the floor much colder than higher in the room, and an uncomfortable cold and "drafty" air will be felt. The colder the outdoor temperature gets the colder the air at the floor level is going to be.

We checked the temperatures in a home that had a gas furnace that was two times larger than it should have been. This home was built in 1945 and had insulation blown in the walls and attic. The windows were replaced with energy efficient ones. The outdoor air temperature was 0 degrees F (-18 degrees C). We measured the air temperature in the room as soon as the furnace blower shut off: it was 86 degrees F (30 degrees C) at the ceiling, 70 degrees F (21 degrees C) five feet (1.5-meters) off the floor, and 63 degrees F (17.2 degrees C) 2 inches (5 centimeters) off the floor. Right before the furnace came on again, the temperature was 83 degrees F (28.4 degrees C) at the ceiling, 68.5 degrees F (20.3 degrees C) at 5 feet (1.5 meters) off the floor, and 58 degrees F (14.5 degrees C) 2 inches (5 centimeters) off the floor. With a heating system like this, the use of a ceiling fan will mix the air better, so the "wasted" overheating of the top of the room (where no one can feel it) is brought back down to the floor. The room will be a lot more comfortable, and allow you to save money by turning the thermostat back a few degrees (while still feeling warmer).

PROPERLY SIZING YOUR GEOTHERMAL HEAT PUMP WILL MAKE A BIG DIFFERENCE

Geothermal heat pumps can be the most comfortable and money saving furnaces you will ever buy, but only if they are correctly sized. An over-sized geothermal will NOT increase your home's comfort over alternative systems. Even worse, an over-sized geothermal (or an undersized earth loop) will run inefficiently, costing you more every month, and wear out sooner than a correctly sized system. This hamstrings a geothermal's benefits.

Over-sizing a geothermal heat pump greatly reduces the running efficiency of a unit. A geothermal heat pump takes about 10 minutes of running to get to 95% of its efficiency. The longer its cycles run, the longer it runs at the high COP's it is advertised at and capable of. A correctly sized unit that runs longer cycles will save you more money on your heating and cooling bills because high COP's mean you get more heat for your money.

Over-sizing a unit and causing it to run shorter cycles will also reduce its life. It's hard to believe, but even just a 10 percent over-sized geothermal heat pump will last about half as long as one that is sized correctly. Unfortunately, we know this is true because we have been asked to consult and troubleshoot on many geothermal heat pump systems that were put in by others. Bottom line is, when a geothermal is oversized and short-cycles itself to death, the only thing that can be done is replace it.

Rather than allow this to happen, always have your system matched to the home it is heating and cooling with a correct heat loss/gain and design! Then your geothermal will live up to the potential inherent in these systems of low energy bills, high comfort, and long life.

RULES OF THUMB FOR SIZING ARE OUT!!

We don't use rules of thumb for geothermal heat pump installations.

People install geothermal heat pumps to save the most money possible. If the system is sized and designed accurately, it will save thousands of dollars on heating and cooling bills. If designers use rules of thumb and guess, and hope they are getting it right, then most people will get cheated out of performance and bill savings. Some people will get cheated out of a little bit of savings, and some people will get cheated out of a lot. Using rules of thumb is not good enough.

Installing a geothermal heat pump is more difficult and costly than installing a fossil fuel furnace, so it's important for you to get all the savings you're expecting to get. Anyone who is going to guess, and depend on luck for the savings, should just save themselves the extra work and expense of installing a geothermal heat pump, and stick with a fossil fuel furnace.

Not convinced? Professors Jeffrey Spitler and James Cullin at the School of Mechanical and Aerospace Engineering at Oklahoma State University lay out the facts:

Misconceptions Regarding Design of Ground-source Heat Pump Systems

"The use of rules-of-thumb for [earth loop] design length remains common in practice, and often leads to oversized, expensive systems or undersized failures. In reality, there are no generally-applicable rules-of-thumb that cover the diverse range of buildings and ground heat exchanger scenarios. Procedures based on building and ground heat exchanger simulation, accompanied by measurement of ground thermal properties will lead to successful designs. Though these procedures are more time-consuming in the design phase, they are a necessary prerequisite to successful, efficient GSHP systems."

CLOSED EARTH LOOPS

A geothermal closed loop has 3 basic parts: the pipe that is buried in the ground, the fluid within the pipes that is used to carry heat, and the pump(s) that keeps the fluid moving.

The pipe used for closed loops is almost always high-density polyethylene (HDPE), PEX-a is also used but a special kind of fitting must be used. The pipe is available in various diameters and lengths. The number of loops of pipe and how deep they are buried are determined by the geographical location, the geological composition, and the amount of BTUH's the geothermal heat pump must transfer.

The fluid will be water or a water antifreeze mixture. Different antifreezes have different properties, and care must be used when choosing among them. For example, you could end up with an antifreeze that gels in your system, or one that is illegal to use in your area!

The loop pump(s) must be specifically calculated for each job. Tying back into what has already been discussed, the number of loops the earth loop is designed with, the diameter of the pipe in each loop, the length of the pipe in each loop, and the type of liquid heat exchanger in the geothermal heat pump, are the major factors in sizing the loop pump(s).

By giving careful consideration to each of these areas, the end result will be a correctly designed earth loop that will provide proper heat exchange to your geothermal heat pump project.

HORIZONTAL EARTH LOOPS

Horizontal earth loops are lengths of geothermal loop piping that are buried 4 to 12 feet in the ground, in horizontally dug or trenched ditches. They are the most common type of loop used for DIY geothermal heat pump installations, because many people either have access to excavators, or can rent trenchers locally. Also, much of the "cost" of these loops is the labor (of hauling pipe around, moving dirt, etc.), which means they are DIY-friendly.

We only recommend that a single pipe is put in a trench. This is because horizontal earth loops absorb heat from underneath the pipe. If multiple pipes are stacked vertically into one trench, only the bottom pipe will do any significant heat absorption (the other pipes just fight with this pipe over who gets the heat). If you are short on land, it is better to do a slinky loop than put multiple pipes in a trench.

The smaller the area that you install the earth loop in, the lower the temperature of the earth loop fluid will drop in the heating season, and the higher it will go in the cooling season. This causes greatly reduced efficiency, in both seasons, and possible shutdown of the geothermal heat pump in the heating season.

HORIZONTAL BORES

Horizontal bores are like vertical bores, except drilled horizontally. This does not do as much damage to the landscaping as installing the other types of horizontal loops, but they have slightly higher operating costs than vertical bores. Be sure to pull a tracer with the pipes. Some states also require the earth loop pipes be marked for GPS locating.

Horizontal loop; single pipe in a trench each pipe spaced 8-feet apart.

This earth loop was excavated using a backhoe.

Horizontal loop; single pipe in a trench each pipe spaced 8-feet apart.

This earth loop was excavated using a backhoe.

Horizontal loop; single pipe in a trench each pipe spaced 8-feet apart.

This earth loop was excavated using a backhoe.

Core drilling the basement wall

SLINKY LOOPS

Slinky loops are made of geothermal pipe that is coiled into many loops and buried in horizontally dug trenches in the earth. This loop requires more pipe, but less land (and excavating cost) than a single pipe trenched horizontal loop. It will often require a larger diameter pipe, and thus more antifreeze (making it more costly), than a single-pipe, trenched horizontal loop.

NAEEM'S SLINKY EARTH LOOP INSTALLATION

Naeem made his own slinky loops by following our instructions. He used 800-foot lengths of 1-inch HDPE DR-11 pipe.

Before digging, he made sure the slinkies would fit where he was planning to dig his trenches.

The trenches are 100 feet away from the basement wall, so some of the pipe will be buried as a straight length.

Normally we don't use slinky loops, because you must install more pipe for the same heat transfer, but Naeem has saturated soil and a lot of underground water flow, so a slinky loop worked for his conditions.

Beginning the excavation.

The trenches are 4-feet deep and 3-feet wide.

These slinky earth loops service a 3-ton geothermal heat pump. Since they are buried in saturated soil, we only needed 85-lineal feet of trench per slinky loop, times 4 trenches. If there had been enough room for 115-foot trenches, we could have gotten away with installing 3 slinky loops.

Backfilling the first trench.

If the soil had been damp clay, we would have needed 3 slinky loops, each in 185 feet long trenches, to transfer the same amount of heat.

Ready to lay the slinkies in the trenches.

The straight runs of pipe from the slinkies will be buried in narrower trenches, but these haven't been dug yet.

Grading the yard over 2 of the buried slinky loops.

Slinky going under a tree root.

Here's the narrow trench for the return lines coming from the third and fourth slinky loops.

Continuing the trenches for the straight supply and return lines to the house.

You may be wondering why we say this is saturated soil, when there is obviously no water to be seen in these trenches.

This soil is normally so soaked with water that it seeps out of the ground on the hill down to the lake. Trying to dig under those conditions would be a disaster, so Naeem waited for the end of August, the driest time of the year, to put his loops in. That's why there's no water laying in the trenches.

The 4 supply lines coming from the basement. The supply line trench and the return line trench are 8 feet apart.

Each supply line penetrates the basement wall through a 2-inch PVC wall sleeve, sealed with a rubber seal. The return lines are installed the same way, in another trench 8 feet away.

The earth loop is buried and the yard is graded.

We're almost done and I wish I had another one to install!

POND LOOPS

Pond loops are coils of geothermal loop pipe attached to a framework and sunk into the water of a pond (or lake). Pond loops are always closed loops. You cannot run surface water directly though your geothermal heat pump's heat exchanger because it will foul your heat exchanger (with algae, other bacteria, scaling, etc.).

Not every pond can be used for a pond loop. The pond/lake must be large enough in volume, and deep enough that the entire loop, when installed, will have 8 feet of water above it.

The pond can't be too far away from your home. If the length of pipe going to the pond would be long enough to satisfy the entire loop length requirement, then the pond isn't useful for your loop. Also, if the pond is far enough away to require fusion welding of the pipe, then DIYers shouldn't use the pond.

Pond loops cut down on loop costs considerably. A pond loop uses a lot less pipe than a horizontal earth loop to transfer the same amount of heat - and since they use less pipe, they also use less antifreeze. Pond loops also have much smaller excavation costs, because the only digging you do is to get the earth loop piping from your home to the pond and back again.

VERTICAL BORE LOOPS

Vertical bore loops consist of HDPE pipe that runs in holes drilled vertically into the earth. Installing these loops usually requires a well drilling machine, and is thus not a Do It Yourself job. The drilling contractor will install the pipe in the bore holes and use thermally-conductive grout to seal off the aquifers, and maintain good thermal conductivity in the loop. Short sections of horizontal trench are dug to connect the borehole loops to each other, and to bring the loop into the house.

Vertical bores are the most expensive type of loop to install. They require the least amount of land, and cause the least yard damage, during installation.

APR 20 2006

APR 20 2006

USING WELL WATER FOR AN OPEN LOOP

An open loop is an earth loop that uses the water from a well to heat and cool your home. The water is pumped from the well through the geothermal heat pump's water-to-refrigerant heat exchanger coil and then returned to the earth. In the cooling season it rejects heat from your home into the water, and in the heating season it absorbs heat from the water into your home.

Show a well pump system here

Open loops were common 25 years ago, but since then, the closed earth loop has taken over the lead. This is partly due to the major improvements in closed loop technology, and partly because of environmental concerns in some areas. The EPA has been considering outlawing pump and dump open loop geothermal heat pumps, but using open loops that return the water back into the earth via an injection well will most likely continue to be acceptable practice. You can still use an open loop with your geothermal heat pump, if it is allowed in your county and state.

For a well to be used as an open loop, it must meet three criteria: well capacity, water chemistry, and pumping power costs.

80

WELL CAPACITY

Many areas of the country do not have enough water in the earth to satisfy the water flow rates of a geothermal heat pump. The amount of water required for the operation of a geothermal heat pump on an open loop is 1.5 gallons per minute, per ton of capacity. For example, if you need a 3-ton geothermal heat pump, your water requirements would be 4.5 gallons per minute. The temperature of the water may increase the flow requirements. During heating mode, if the water temperature is lower than 41 degrees F, the flow must be increased until the leaving water temperature stays above the freeze protection settings. In warmer climates, during cooling mode, the water flow rates may need to be increased so the geothermal heat pump's efficiency will be acceptable.

The annual amount of water used by a 3-ton geothermal heat pump is about one million gallons a year. This is a very large amount of water, but since geothermal heat pumps don't change the water quality, only water temperature, all of the water used by a geothermal heat pump can be safely returned back to the earth without contaminating the ground/aquifers or wasting any water.

If you are going to install the system on an existing well that already supplies water for domestic use, then the well will have to have sufficient capacity to meet the needs of both the geothermal heat pump and the domestic water usage.

WATER CHEMISTRY

There are many factors that will determine if your water's chemistry is satisfactory. Bad water chemistry will either scale the heat exchanger excessively, which is a nuisance, or dissolve the water-to-refrigerant heat exchanger out of your geothermal heat pump, which is a disaster. The mineral profile, pH, and temperature of the water must all be within the correct ranges for the well water to be compatible with a geothermal heat pump's water coil heat exchanger.

Water that has a pH level too high or too low will dissolve the heat exchanger, and can't be used. Also, the warmer the water, the faster an acid or alkaline will dissolve the heat exchanger, and the smaller the range of acceptable pH's. In the south, ground water temperatures are higher, and more time is spent in cooling mode (when the geothermal heat pump is adding heat to the water), so southern open loops must have more neutral pH than northern open loops.

Some types and amounts of dissolved minerals will scale the heat exchanger over time. Scaled heat exchangers can be cleaned, but it is a nuisance and an added cost. The frequency that cleaning is needed will determine whether or not this type of well is usable. Most of the time it would be better to use a closed earth loop and forget about the inconvenience of cleaning the water coil, no matter how many years it would last between cleanings.

WELL WATER PUMPING COSTS

The costs of the well pump will also affect whether or not a well should be used for an open loop. The main costs of pumping well water are how far it must be lifted up against gravity, and the type of motor used in the pump. Most electric motors that are used on submersible water well pumps are permanent split capacitor (PSC). Until a few years ago, that was the highest efficiency motor available for residential use, but now there are constant pressure/ECM well pumps available. The ECM (electronically commutated motor) has a very high efficiency and will cut the cost of pumping water by 60 percent, compared to a PSC motor. The ECM motor also makes the pump variable speed, which means your water pressure will be constant. The downside to the constant pressure/ECM pump is the initial cost. Depending on the model, the pump may cost $1200.00 to $1800.00.

GEOTHERMAL HEAT PUMP EARTH LOOP ANTIFREEZE

In the majority of the USA and Canada, residential geothermal heat pump earth loops must be protected from freezing. Many people that have contacted us have asked, "If I bury my earth loop deeper, will I still need to use antifreeze to protect my earth loop from freezing?" This question tells us that there is a misconception of why the earth loop is subject to freezing. If your earth loop is installed above the frost line, yes it will freeze, but even if you install your earth loop below the frost line, the fluid may still freeze. Freezing in an earth loop is caused by the geothermal heat pump taking heat from the loop fluid, not the winter air temperatures.

The more undersized a loop, the longer it takes for the earth to recharge the heat in it, as the geothermal heat pump removes the heat. We have been asked, "Can't we just design bigger earth loops, so the earth loop fluid doesn't drop below the freezing temperature?" Sure, this can keep the earth loop above freezing. However, earth loops must be designed for a reasonable payback period. If we design to keep earth loop fluid above freezing, it can require twice as much pipe in the earth loop in many geographical locations, and in some places it requires 10 times as much pipe. Since it is not cost-effective to use this much pipe, we use antifreeze to keep the fluid flowing below 32° F, instead.

What types of antifreeze can be used in geothermal earth loops?

METHANOL (WOOD ALCOHOL, METHYL ALCOHOL)

For years methanol was the choice for many geothermal heat pump installers. Methanol works very well because it is less expensive to buy, and flows better (has low viscosity) down to 15° F than other types of antifreeze. It also has good heat transfer ability (it holds acceptable amounts of heat, compared to water). However, methanol is extremely poisonous to humans and other animals. It evaporates quickly, and can asphyxiate a person if all of the safety precautions are not followed. Methanol is also highly flammable, and can cause an explosion.

Because of methanol's toxicity, some states in the USA have outlawed its use in earth loops buried deeper than 20 feet, and other states have outlawed its use in all earth loops, to protect the groundwater* if the loop should leak. Methanol is not our preferred antifreeze, because of the danger of fire and explosion when working with it, and because of the restrictions that have been imposed on it by many local and state health codes.

*However, research has shown that methanol does biodegrade rapidly in the natural environment; even deep below the ground.

ETHANOL (DENATURED ALCOHOL, ETHYL ALCOHOL)

Ethanol is another antifreeze that has been used for geothermal earth loops. It has similar characteristics to methanol as an antifreeze: it flows well, has good heat transfer, and good freeze protection at 15° F. Ethanol is also very flammable and can cause explosions and asphyxiation.

Pure ethanol (the type of alcohol that people drink) is not as toxic as methanol, but pure ethanol is too expensive to be used as an antifreeze, so denatured ethanol is used instead. These denaturing agents *are* generally very toxic. Ethanol can be denatured with methanol, pine-based solvent, gasoline, rubbing alcohol, or other such chemicals. Ethanol that is denatured with petroleum-based products will dissolve earth loop piping, and can't be used for earth loop antifreeze.

Some brands of ethanol-antifreeze are available that have been specifically designed for use in geothermal earth loops. They contain denaturing agents that are safe for geothermal pipe. We have used a brand called "Geosafe" (safe for pipes, but still dangerous for people), but there are others available.

ETHYLENE GLYCOL (CAR ANTIFREEZE)

Ethylene Glycol is a very poisonous antifreeze. It also becomes very thick and flows badly (has high viscosity) at temperatures below 35° F, and has fairly low heat transfer ability. Also, most states have prohibited its use, because of the dangers of contaminating and poisoning groundwater. We therefore do not recommend ethylene glycol be used in an earth loop.

PROPYLENE GLYCOL

Propylene glycol is a non-toxic antifreeze. It is used in the food preparation industry and is considered safe*. However, it has some viscosity problems that limit its use as an antifreeze for flowing fluids. Special care must be taken when designing and calculating earth loops that will use propylene glycol, or the fluid flow will be too fast in the summer, and too slow in the winter.

Propylene glycol's low toxicity makes it the only earth loop antifreeze allowed in many states. We recommend this antifreeze, but only for earth loops that have been properly sized and designed by a professional earth loop designer.

*Propylene glycol is [toxic to cats](), however, so don't let them drink it.

CALCIUM CHLORIDE

Calcium Chloride performs well as an antifreeze in earth loops. It is non-toxic but extremely corrosive. If this type of antifreeze is used, the geothermal heat pump's water coil heat exchanger must be cupro-nickle. Any metal fittings, pipes, or pumps must be made from brass, or better material, for corrosion protection. We also recommend this antifreeze.

WHAT IS INSULATION R-VALUE?

Insulation is rated with an R-value number. This number equals the thermal resistance of a material to allow heat to transfer from one side of the material to the other. When considering insulation for your home, materials with higher R value ratings will be better at slowing the transfer of heat from your home. In the winter these values slow the heat from moving from the inside of your home to the cold outdoor air, and in the summer these insulation values will slow the heat from moving from the hot outside air to the cooler inside air in your home.

R-1 has a low resistance to heat transfer. This means heat will move through it fast and thereby cause your furnace and air conditioner to use more energy to keep your home warm in the winter and cool in the summer. R-2 means 1/2 the heat loss (or gain) of R-1. R-15 means only 1/15th as much heat as R-1 will get through, and so on. Higher R values thereby cause your furnace and air conditioner or geothermal heat pump system to use less energy to keep your home warm in the winter and cool in the summer.

BEST INSULATION R-VALUE PER DOLLAR: STOP THE LARGEST LOSSES/LEAKS FIRST

What is R-value?

The way a home or business building is insulated has a huge impact on the heat loss/gain.

The best places to insulate will always be the areas where the most heat per square foot is leaving (in the heating season) or entering (in the cooling season).

We have been trained to believe that the best place to insulate is the attic or ceiling no matter what. This saying started years ago when homes were not insulated at all. Installing insulation in the attic would save the most money, because the air at the ceiling is warmer (since heat rises), and insulating the attic was a lot easier than insulating any other part of the house.

This saying was true when insulating of homes first began, but it has mislead most people of today. If you had a home without any insulation, and you had money to insulate one area of it to an R-5, of course the attic would be the most important place to spend the money. If you insulate your attic to an R-30 but do not insulate anywhere else you will have wasted 2/3 of the insulation. It would be like insulating the attic to R-30 and then removing a couple of windows, and leaving a couple of big holes in the walls.

For years installers did not even consider the basement area in the heat loss calculations, but now installers are starting to realize the considerable loss there is though masonry walls (cement block, poured concrete, brick), even if your basement temperature is kept at 55 to 60 degrees. Your basement walls MUST be insulated to benefit the most from a geothermal heat pump.

It is true that if your home had no insulation installed in it at all, the basement would have the lowest heat loss of any floor that is above the ground level. But, when you have even some insulation in your other walls and ceilings, heat loss calculations then show the basement is now the place of largest loss.

I have been told by many people that the earth is warm and for this reason the basement should not be insulated. Sure the earth is warm: this is where a geothermal heat pump transfers the heat from. The problem is, when there is only a little thermal resistance to the heat transfer through the wall (R-value), then even with a small temperature difference there will be mega BTUH's (BTU's per hour) transferred back into the earth from the basement. It is silly to pay for power to move heat out of the earth and have most of it go right back into the earth through badly/not insulated basement walls.

Some people tell us they don't even heat their basement, so it doesn't need insulation. The facts are this: If you live in a cold climate and you did not heat your basement, it would eventually freeze. Your basement is above freezing in the winter because it absorbs heat from your first floor (and a little bit from the ground around it), which means you are heating it.

In the deep south it is usually more beneficial not to install insulation in the basement because of the cooling benefit of the cooler earth temperatures. Because the earth's temperature is 60 to 75 degrees, there will be a much lower impact on heat loss. But it is the heat loss/gain calculation that accurately tells you whether your basement is gaining or losing heat and how much, and that is how you determine where best to spend your money on insulation.

An example of how large a basement's heat loss can be: In a home in an area where the low temperature can be 10 degrees F, a masonry basement wall (R-1.83) will have an overall average temperature of 32 degrees F. Even if your basement temperature is at the lowest allowable temperature (if the geothermal heat pump is installed there) of 55 degrees F, the heat loss will still be about 22,000BTUH. If the basement walls are insulated to an R-10 the BTUH heat loss would be 3,782... WOW look at the difference!!! A ton and a half smaller furnace can be installed!

WHAT IS A DESUPERHEATER?

A desuperheater is a secondary heat exchanger that transfers heat from the earth in the winter, and from your home in the summer, into your domestic hot water tank. The desuperheater is part of the geothermal heat pump's domestic hot water generating system (HWG).

Hot water generators only heat domestic water when your geothermal heat pump is heating or cooling your home. The hot water generator's water circulating pump moves the cold water from the bottom of your hot water tank through a water pipe to the desuperheater itself, where the water is heated by the heat that has been transferred from the earth when you are heating your home, and from the inside of your home when you are cooling your home. The heated water is then circulated back into your hot water tank.

Heating water with a geothermal heat pump's hot water generator costs about 80 percent less in a heating dominated climate, and up to 95 percent less in a cooling dominated climate, than if you heat your domestic water with an electric, oil, or propane fired hot water heater. At the present cost of natural gas the savings is about 15 percent less - not as drastic a savings as the others, but saving on your bills is still saving on your bills!

In the summer heating your domestic water is "almost" free. This is because the heat that is being removed from your home is transferred into your hot water tank. Since this heat from inside of your home was going to be rejected (thrown away) into the earth anyway, putting it into your hot water tank instead is free. The only cost for the summer water heating is the small cost of running the circulating pump that moves the water.

SINGLE STAGE, TWO-STAGE, AND VARIABLE SPEED GEOTHERMAL HEAT PUMPS

What is staging, and how could it be so important?

WHAT IS SINGLE-STAGE?

Not so many years ago, geothermal heat pumps, as well as other heat pumps, were only manufactured as single-stage units. When a single-stage compressor starts, it produces a constant amount of heating or cooling until the compressor shuts off. Since units are sized for the majority of the heating need (or all of the cooling need), they only run for brief periods at a time at the beginning of the heating or cooling seasons. As the season progresses, and more heating or cooling is needed, the compressor runs for longer periods of time to keep the home comfortable. At the peak of the heating season, the compressor will run very long cycles, and even continuously when outdoor temperatures are the most extreme.

When compressors operate in very long cycles, or better yet, continuously, the efficiency of the heating and cooling equipment is at its highest. It seems like running continuously would wear out a compressor faster than letting it cycle, but the facts are just the opposite. When a compressor starts up, it takes a great amount of force to get it to start rotating. For a split second, this extreme force causes the metal surfaces to rub slightly. The wear is minute, but it adds up: a million tiny wears will finally wear it out. And while a modern compressor can start up a million times, short-cycling uses up those starts in a few years.

Short-cycling was an ever-present problem for single-stage geothermal heat pumps. There was a huge need for a system that could run for long periods of time, whether it was 40°F, or 20°F outside. The solution was a geothermal heat pump that could run in different sizes, or stages.

THE FIRST ATTEMPT AT STAGING

Years ago, if a customer could afford it, we would install two smaller geothermal heat pumps and operate one of them on first stage, and the other one for second stage. This type of installation was very costly and since geothermal heat pumps are not inexpensive, it was just not cost effective for most installations.

AN ATTEMPT AT MAKING A TWO-STAGE

Since the need was so great, engineers came up with a unit that had two complete refrigeration circuits in one geothermal heat pump furnace. These systems were designed with two compressors, and two of everything else that makes a system work, except the cabinet, and the blower motor. This unit was still costly, and even though it solved many of the cycling problems, it was just not the answer that the geothermal heat pump industry needed.

THE TWO-STAGE COMPRESSOR

Finally the two-stage Copeland Scroll Ultratech compressor was invented. This compressor was the answer that the industry had been looking for. It is able to operate at 67 percent of its capacity, and while it's running, shift to second stage and 100 percent of its capacity. It greatly reduces the short-cycling problem, as long as the equipment has been sized correctly. The Copeland Scroll two-stage compressor has been a revolution for the geothermal heat pump industry, but HVAC contractors must stop oversizing geothermal heat pumps before the short-cycling problem can be eradicated.

AN EVEN BETTER COMPRESSOR

This new compressor is variable speed, and will eliminate the short-cycling problem forever. Since it will be able to operate at whatever capacity is needed, it will almost never shut off, allowing for 50 percent higher efficiency then we have now. This new variable speed compressor is available in geothermal heat pumps NOW.

WATER-TO-WATER VS. WATER-TO-AIR
GEOTHERMAL HEAT PUMPS

If you have been reading about geothermal heat pumps, you may have heard them called water-to-water, or water-to-air. What do water-to-water and water-to-air mean?

WHAT DO THEY DO?

Water-to-water geothermal heat pumps heat and cool water, for hydronic radiant systems, or for dedicated water heating in commercial buildings (for example, washing machines in a laundromat or hospital).

Water-to-air geothermal heat pumps heat and cool air, for forced-air ducting systems. In the United States, most homes use forced-air ducting distribution systems, so water-to-air units are the standard type used.

WHAT PARTS ARE DIFFERENT?

A water-to-water unit has 2 water coils in it. One is connected to the earth loop, and the other is connected to the buffer tank, and then the hydronic distribution piping. The water in your hydronic piping circulates through this coil and is heated or cooled.

A water-to-air unit has 1 water coil, and 1 air coil. The water coil is connected to the earth loop, and the air coil is connected to your air ducting. Air from inside your home is circulated through this air coil to be heated in the winter, or cooled in the summer.

Water Coil + AirCoil =

Water-to-Air
Geothermal Heat Pump

Water Coil +Water Coil =

Water-to-Water
Geothermal Heat Pump

Are There Any Other Differences?

- Water-to-air packaged units include the fan assembly inside the cabinet as standard equipment. However, the circulating pumps for water-to-water units are an additional purchase.

- Water-to-water units may be available in fewer sizes than water-to-air units.
- Water-to-water units when coupled with in-floor radiant heating are more efficient than water-to-air units.

GEOTHERMAL HEAT PUMPS: PACKAGED UNITS VS. SPLIT SYSTEMS

The standard type of geothermal heat pump is a packaged unit. The entire refrigerant circuit, including both heat exchanger coils (water-to-refrigerant/source and air-to-refrigerant/load), is inside one cabinet. The unit is entirely assembled and charged at the factory, so that no on-site refrigerant or brazing work will need to be done. This is the most common type of geothermal heat pump installed.

Split system geothermal heat pumps are separated into two sections: the compressor and the water-to-refrigerant heat exchanger in one cabinet, and the blower assembly and air-to-refrigerant heat exchanger in either an air handler cabinet, or in a fossil fuel furnace. The refrigerant circuits of these components must be connected on-site with brazed copper refrigerant tubing, and then charged with refrigerant.

Split systems have a shorter life, on average, than packaged geothermal heat pumps. Split systems are much more prone to contamination during assembly than packaged units are. It is more difficult, and requires more skill, to braze and charge a refrigerant circuit at a home site than at a factory. Any contamination of the refrigerant circuit causes compressor burn-out and system failure. For this reason, we don't recommend that our customers install split system geothermal heat pump unless it is the only way it can be installed.

CLIMATEMASTER CLIMADRY GEOTHERMAL HEAT PUMP DEHUMIDIFICATION SYSTEM

Many areas of the country have very high humidity, even when temperatures aren't hot enough for air conditioning. This makes your home feel damp, muggy, and uncomfortable, and can even cause mildew. Many people that live in areas like this buy whole house dehumidifiers, but these cost a lot to operate.

The ClimateMaster ClimaDry geothermal heat pump system is great for these areas of the country where the humidity is high, but the outside air temperatures are not hot enough to allow complete dehumidification by the air conditioning system. What is nice about the ClimaDry unit is that it will keep the air in your home very comfortable all year round.

With ClimaDry, your thermostat is able to tell how humid it is in your home. When the humidity is too high, the geothermal heat pump will switch to the ClimaDry dehumidification mode, and run until the air has been dehumidified and made to feel absolutely comfortable. As it is removing the humidity, it keeps the temperature right where you set it, all automatically, and for a fraction of the cost of running a whole house dehumidifier.

If you are sizing your equipment with us, we can give you an estimate for a ClimateMaster ClimaDry geothermal heat pump system, and estimate what this system will cost you to operate.

GEOTHERMAL HEAT PUMP THERMOSTATS: KEEP IT CONSISTENT!

It is always best not to turn your geothermal heat pump's thermostat down and then back up to save energy; you will not save and probably it will cost you a lot of money turning the thermostat temperature up and down. You have probably heard otherwise, that this is a good way to save money on your heating bills, so why should we say this? The simple answer: it works well for fossil fuel furnaces, but it backfires when used with geothermal heat pumps.

A few years back, we had a customer buy a new thermostat and install it with a geothermal heat pump system we installed. The first electric bill they got, after they installed their new thermostat, was $425.00 higher than it should have been.

The new thermostat that they replaced our original thermostat with was a programmable one. Someone else told them they could save a lot more money if they installed a programmable thermostat and programmed it to set back to 62 degrees during the day when they were at work, and back up to 70 at 5:00 in the evening, and down again at 11:00 at night, and then up again so their home would be warm in the morning.

This works well with fossil fuel furnaces, because fossil fuel furnaces are sized way bigger than they need to be to get the job done. They can play "catch up" with temperatures in the house. They can do this without becoming very much more inefficient than they already are. Geothermal heat pumps are extremely efficient when sized correctly. This means, however, that they don't "catch up" as fast as fossil fuel furnaces. This isn't a problem for normal heating or cooling of a home, since we want to live in a pretty steady temperature anyway, but it does mean setting the thermostat back to save money won't work. If you let the temperature drop say 10 degrees, and then want it to come back up those 10 degrees in, say, an hour, the back-up heat (electric resistant or natural gas or fuel oil) has to come on to help the geothermal do it. Backup heat is more expensive than geothermal heat, and the few pennies of savings from turning the geothermal heat down are wiped out with the extra dollars it costs to quickly heat it back up with backup heat.

Our customers ended up heating their home mostly with electric resistant heat, and that cost them $425.00 more than the $150.00 it should have cost to heat their home for that month, a February in Akron, Ohio.

Setting a furnace back is what people with expensive, inefficient fossil fuel furnaces had to resort to, so they could save money somehow. Geothermal heat pumps are already money saving and efficient, and because of their design, don't benefit from the technique of setting temperatures back.

We supply a programmable thermostat with the geothermal heat pump systems we sell, in case a customer likes the temperature in their home to be a few degrees cooler or warmer at certain times of the day or night for their own comfort. They will have the convenience of having the thermostat do the work so they don't have to be running to turn the thermostat up or down whenever they want a little change in temperature. This is not done to save money, only for personal comfort preferences. Even a couple of degrees to catch up will most likely require the use of auxiliary heaters, if it is very cold outside.

The moral of the story is: set the thermostat where you want it, and keep it there.

DUAL FUEL
GEOTHERMAL HEAT PUMP SYSTEMS

Dual fuel geothermal heat pump systems are not as efficient as other types of geothermal heat pumps, because dual fuel systems must turn off the geothermal heat pump's compressor when auxiliary heating is needed. Our experience with these systems is that when it is very cold and the thermostat calls for auxiliary heat, the geothermal heat pump's compressor will shut down, the fossil fuel furnace will fire to warm the home up, and then the system cycles off. Because it is very cold outside, in only a few minutes the thermostat will call for heat again. The geothermal heat pump will start and begin to heat, but by the time it is making heat the home has already dropped in temperature enough that auxiliary heat is called for and the geothermal heat pump's compressor is shut down again.

Using a number of systems to adjust the controls and stop this kind of cycling has resulted in no success. When you use this kind of system, and it gets cold enough outside that you need auxiliary heat, 90 percent of your heating will be from fossil fuel. Heating with fossil fuel as the majority heating source defeats the purpose of the geothermal heat pump.

If you have the space, you can leave your existing fossil fuel furnace and connect a geothermal heat pump next to it. This is different than using a dual fuel split geothermal system because your geothermal heat pump's compressor doesn't have to shut down when auxiliary heating comes on. This means auxiliary heating will come in much smaller doses, and more importantly, that the geothermal heat pump will still do the majority of the heating. There are some controls that need to be installed for the air flow, but typically it is not more complicated than that.

Please note: Dual fuel split geothermal heat pump systems must be installed by a certified HVAC technician. Since they have 2 sections, the refrigerant lines must be connected and the system charged with refrigerant, and this MUST be done by a certified HVAC technician.

ABOUT HEAT DISTRIBUTION

Once we've taken heat out of the earth with the earth loop and squeezed it into higher temperatures with the geothermal heat pump, we must move it throughout your home. Your heat distribution system is how that is done, with **air ducting** or a hydronic system (radiant in-floor heating, baseboard radiators). Air ducting is the most common form of distribution in the USA.

It is extremely important for heat distribution systems to be sized and designed correctly, for your geothermal heat pump to operate efficiently. Most existing hydronic systems are designed for the higher temperatures of a fossil fuel boiler, and need reworked before they will function at all with the efficient, lower water-output temperatures of a water-to-water geothermal heat pump.

Your geothermal heat pump will be matched to the type of heat distribution system that you use (or vice versa): air ducting goes with a water-to-air geothermal heat pump, and hydronic heat distribution goes with a water-to-water geothermal heat pump.

SIZING AND DESIGNING YOUR FORCED AIR DUCTING SYSTEM

We can help you design or evaluate your air ducting for use with a geothermal heat pump, whether you are building a new home, or replacing your existing fossil fuel furnace. Most forced air ducting systems are designed to be used with fossil fuel furnaces, which output air at about 130 – 160° F. A water-to-air geothermal heat pump outputs air at about 95 – 100° F, meaning it is more efficient, but requires more airflow for proper operation. Your air ducting must be able to provide this higher airflow, efficiently, and without being noisy. Additionally, the duct design should allow the proper amount of air to reach each area of your home.

The total air flow through the geothermal heat pump/air ducting must be at the geothermal manufacturer's rated CFM (cubic feet per minute), and at the rated static pressure. If the air ducting is too small, the air will move too slowly, causing inefficient geothermal operation. Simply increasing blower speed will not solve the problem, since the extra blower energy use will lower the system's efficiency. The only way to get the correct amount of air flow, with acceptable energy use, is to have large enough air ducting.

The air ducting system should also be quiet. The blower noise, or any noise that the geothermal heat pump makes, must not travel through the air ducting system. Using vibration dampers on the supply and return air ducts or plenums, and lining these areas for 5 feet in each direction with an approved sound deadening material, will usually prevent this. Also, room air registers should be sized large enough to let the air flow quietly through them. You shouldn't be able to hear your geothermal heat pump running; you should only know it is operating by the very gentle movements of your curtains.

Finally, the air ducting must provide the correct air flow to each room. Air flow is affected by the air pressure in the ducting, which is determined by the length and size of the duct, and the speed of the blower motor. Also, every room should have a return air register, or else air flow will be interrupted when the room's door is shut. For the best control of room air flow, an automatic duct damper system should be used. Automatic duct dampers open or close air ducts as a room's air flow requirements change, according to the room's thermostat. They allow each room to be kept at a few degrees warmer or a few degrees cooler than the main system setting, if the occupant desires it. They also automatically adjust for the effects of the sun shining on one side of the home, or different amounts of air infiltration from the wind, so that each room stays at the desired temperature.

Air ducts that are installed in unconditioned spaces will need to be insulated to the most recent building codes for their locale. Most new insulation codes require that air ducting in unconditioned spaces be insulated to at least R-11, and be air tight. If air ducting is installed in attic spaces, those spaces must be ventilated so the air temperature will not rise higher than a few degrees above outside temperature. If you are going to install the geothermal heat pump unit itself in any of these unconditioned spaces, you must make sure it is rated for it.

SIZING AND DESIGNING YOUR GEOTHERMAL HEAT PUMP

Your system will need to be sized and designed correctly, and this is done by calculating your home's heat loss and heat gain.

The payback for a correctly designed system will take no time at all. It's not just the cost of the heat loss/gain calculation that will be paid back, a correctly sized and designed system will pay for itself over and over.

People call us every day telling us of problems that they are having with their geothermal heat pump systems that they had installed, and we have helped a lot of them fix their geothermal heat pump problems, but sadly to say, many systems have been installed so poorly that they cannot be corrected without making major changes to it.

The savings on your heating and cooling bills, the comfort you will have, and how long of a life your geothermal heat pump will have, hinges on it being designed, sized, and installed correctly. A few hundred dollars spent now could save you many thousands of dollars because of a bad job.

The first thing that needs to be done is a correct heat loss/gain calculation for your home so we can size your geothermal heat pump for you. Once we know the size of your geothermal heat pump we can size and design your earth loop for you.

You will need to add up all of the square footage, including the basement, and then use the drop down menu on our website to select the square footage and then you can pay for the unit sizing.

 Once you have paid for your geothermal heat pump sizing we will send you and email requesting the information that we need, so we can calculate the heat loss/gain for your home. The information needed is on our website geojerry.com . If you have PDF files of your blue prints, be sure that all of the information that we have requested on our website is on the plans and then email them to us (you will have our email address by then). Once you send us the information we will look it over and let you know if we have everything we need. Once we receive all of the information we have requested from you it will take one to two weeks to send you the heat loss/gain for your home and what size geothermal heat pump you will need to install.

DESIGNING YOUR CLOSED EARTH LOOP

A geothermal heat pump's earth loop installation is not just a matter of burying some pipes and hooking these pipes to your geothermal heat pump. Your earth loop must be matched to the home, and the geothermal heat pump, that it is coupled to. So, the first step in sizing an earth loop is having a correct heat loss/gain calculation done. We need to know how many BTUHs your earth loop needs to supply to your geothermal heat pump, so it can heat your home correctly. This also tells us what the flow rate through your earth loop will have to be.

Next, we need to know about the soil that the earth loop will be buried in. How fast can it transfer heat (the rate of thermal diffusivity) into, or out of, each lineal foot of the earth loop? To accurately determine your soil's thermal diffusivity, we need to know what your soil conditions are, including the type/s of soil (clay, sand, gravel, silt, top soil, and many others), and the moisture content of the soil. You will need to dig a couple of test holes so a soil analysis can be done. You can either do this soil analysis yourself, with our instruction, or you can hire a soils engineer.

The thermal diffusivity of the soil, along with the BTUH transfer needed by the geothermal heat pump, will together determine how many lineal feet of pipe (of a certain diameter) will need to be installed in the ground. For example, for horizontal loops: moist clay soil will usually need 600 lineal feet of pipe per ton, very dry soil needs at least 1200 lineal feet of pipe per ton, sometimes more, and saturated soil (where water seeps into the trench as you dig) needs 425 lineal feet of pipe per ton.

Once we know how many feet of pipe the earth loop will need, we must calculate its total pressure loss. This includes the friction inside the pipe, the manifolds, the pipe and fittings that will be inside the home, the hose kit, and the geothermal heat pump's water-to-refrigerant heat exchanger (some have much larger pressure losses than others). We also have to add the friction loss caused by antifreeze viscosity, and the friction loss at the lowest temperature in the heating season. Once we know the total pressure loss, we must adjust the loop (for example, by using smaller diameter parallel loops, or different pipe and fittings inside the home) if the pressure loss would cause high pumping costs.

Designing the earth loop is a balance between how much power the circulating pump will use because of the diameter of the pipe, and how much the pipe will cost because of its diameter. If the diameter of the pipe is small, you will save money on the cost of the pipe; but since the pipe diameter is small, the pumping power needed will be high, and this will greatly increase your energy bills. On the other hand, if the diameter of the pipe is large enough so that the smallest pump can be used, the energy cost that the very small pump saves will never pay for the high cost of the larger diameter pipe.

It is equally important for the geothermal heat pump, earth loop, and air ducting (or radiant heat piping) to be designed correctly.

WHAT WE NEED FROM YOU SO WE CAN SIZE AND DESIGN YOUR HORIZONTAL EARTH LOOP

As I said in the previous chapter sizing and designing your system is an absolute necessity and that goes for the earth loop also.

If you had us size your geothermal heat pump for you we will also size and design your horizontal earth loop for no additional charge.

We will need a few things from you so we can design your earth loop.

We need a sketch of your property (plat map) showing your property lines, where your home sets, and where you want to install the earth loop including the dimensions, but it doesn't need to be drawn to scale. On the drawing show any trees, buried lines for; electric, cable, water, sewer, well, septic, or any other thing that might be in the area of the proposed earth loop. Also be sure that there aren't any easements where you are going to install the earth loop. If you have wetter soil somewhere on your property that's close to your house then that area would be a better choice for your earth loop installation.

HOW TO TEST YOUR SOIL

The soil conditions for your lot must be figured out by digging test holes where the earth loop will be installed. There is no other way to figure it out. Every horizontal loop installation must have soil samples taken or we will only be guessing what size earth loop to design. Most geothermal heat pump contractors guess, and that is one of the reasons that there is so much trouble in the industry. Since geothermal heat pumps are installed for energy (and money) saving, there cannot be any guessing about sizing and designing any part of it.

Most people these days have a video camera, so you can easily make a video while you are doing a soils test and email the video to us. People send information of their soil testing, their furnace rooms, the areas where the air ducting will be installed, their windows, and many other details of their home so we can correctly calculate their geothermal system for them.

THE TEST

Before you do any digging or testing, be sure that it has not rained for at least 24 hours, because sometimes surface water will run into the hole and then the testing results will be inaccurate.

Your earth loop designer will tell you how deep to dig your test holes. You should be able to reach a depth of about 4 or 5 feet with a pair of post-hole diggers, but if you must test deeper than this, you'll need an excavator to dig the holes for you.

You will need to dig at least 2 test holes: one in your front yard and one in your back yard. You may need to dig more if your earth loop designer requires it. You must do the test as soon as the dirt is removed from the trench, or as soon as you dig your last shovel of soil out of the hole. If you have an excavator dig for you, have them dig down to the specified depth, and then have them bring up a scoop of dirt for you to test. NEVER GET DOWN INTO AN EXCAVATED TRENCH THAT IS MORE THAN 4-FEET DEEP BECAUSE IT MAY CAVE IN ON YOU AND KILL YOU.

Have someone take a video of you doing all the tests. Be sure to say where the test hole was dug as you start the test (as in front yard, back yard, side yard, top of the hill, bottom of the hill, etc.).

Dig down to your required depth, and immediately bring up a shovel of soil. Take a handful of soil and close your fist on it tight. Then open your fist so I can see if the soil packs in your hand. Next take a pen or something like it and see how far the pen will push into the soil before it falls apart. Do this test 3 times in a row, pulling a new sample of soil out of the hole for each test.

The purpose of the last test is to tell us how much water is in the soil. Wait 1 hour and look into the hole. If you look into the hole after an hour and see any water standing in it, it means the soil is saturated with water, and that is the best we can hope for. You won't have to bring any more soil out of the hole; just tell us where the hole is and that you saw water in it. If you don't see any water in the hole, bring another scoop of soil out and do one more test in the same way you did the other tests, videoing it also.

OR YOU CAN SEND US THE SOIL SAMPLES

If you don't have a video camera, or you'd rather not do the tests yourself, you can send your soil samples to us and we will analyze them. Dig the test holes in your front and back yard, and have freezer bags ready. Once you are at the 4-foot depth take about 2 cups of soil and put it into a freezer bag, and triple bag it. Try to push out as much air as you can. Mark the bags "First Test Sample" and whether it's from the front or back yard (or wherever it's from).

Wait 1 hour and see if there is any water in the holes, because if you have saturated soil, we won't need the soil sent to us. If you don't see water, then take another soil sample from each hole, and put them into separate triple-bagged freezer bags. Mark them "Sample taken after one hour" and where it's from. Put all the freezer bags into a box and contact us for our shipping address.

About the Author

After seeing a demonstration at a local fair in 1983, Jerry Scherer installed his first geothermal heat pump system in his new home. Being fascinated at the concept of transferring heat from the earth into a building, he started installing them for a living.

In 2003 he started selling them on eBay with DIY installation instructions. After a few months of selling them on eBay he realized that contractors were not sizing and designing systems correctly which was and still is causing many people to get geothermal heat pump systems that do not operate correctly.

Now he mainly sizes and designs systems for people in North America and in a few in other parts of the world.

His goal is to educate the potential geothermal heat pump customers so they can get a geothermal heat pump installed, either by doing it themselves, or by having a contractor install it for them.

He has been installing and troubleshooting geothermal heat pump systems for 30 years.

www.ingramcontent.com/pod-product-compliance
Lightning Source LLC
Chambersburg PA
CBHW050719180526
45159CB00003B/1071